Rogério Tomanini

Sampling evaluation of H. Pylori virulence factors

Rogério Tomanini

Sampling evaluation of H. Pylori virulence factors

SAMPLE EVALUATION OF H. pylori VIRULENCE FACTORS IN DIFFERENT DISEASES OF THE GASTROINTESTINAL TRACT

ScienciaScripts

Imprint

Any brand names and product names mentioned in this book are subject to trademark, brand or patent protection and are trademarks or registered trademarks of their respective holders. The use of brand names, product names, common names, trade names, product descriptions etc. even without a particular marking in this work is in no way to be construed to mean that such names may be regarded as unrestricted in respect of trademark and brand protection legislation and could thus be used by anyone.

Cover image: www.ingimage.com

This book is a translation from the original published under ISBN 978-613-9-66276-0.

Publisher:
Sciencia Scripts
is a trademark of
Dodo Books Indian Ocean Ltd. and OmniScriptum S.R.L publishing group

120 High Road, East Finchley, London, N2 9ED, United Kingdom
Str. Armeneasca 28/1, office 1, Chisinau MD-2012, Republic of Moldova, Europe
Printed at: see last page
ISBN: 978-620-7-99419-9

I dedicate this work first of all to God, because without him nothing would be possible, to my mother Eledi, my father Vagner and my beloved wife Leniza, who with great wisdom and patience helped me and encouraged me to overcome another stage of my journey.

I thank the Master of Masters Jesus for being the beacon that illuminates my life;

To all the teachers on the biological sciences course, especially Prof Adriana Bertolini, who with professionalism and mastery guided me to produce this work;

To Prof Dr Marcelo Lima Ribeiro for believing in my work and giving me a scientific initiation in his laboratory.

I would like to thank all you exemplary teachers.

For wisdom is better than jewellery, and of all the things
you desire nothing can compare with it.

(Proverbs 8:28)

SUMMARY

According to studies in various countries, the high incidence of the development of diseases of the human digestive system is directly related to infection by the bacterium **Helicobacter pylori**, which causes an inflammatory process in the stomach mucosa followed by an increase in the production of cytokines, giving rise to various pathologies such as: dyspepsia without ulceration, reflux oesophagitis, erosive gastritis, chronic gastritis, gastric peptic ulcer, duodenal peptic ulcer and gastric cancer. In this process, oxygen and/or nitrogen free radicals are generated, causing oxidative damage to DNA. It has been found that the most virulent strains produce greater damage to the DNA of gastric cells, in addition to inducing apoptosis and also stimulating cell multiplication, thereby producing alterations in the **turnover** system. In this study, the **cagA** and **vacA m1/m2-s1/s2** virulence factors of **Helicobacter pylori** were assessed in samples from 45 patients from the São Francisco University Hospital in Bragança Paulista / SP, from June 2011 to January 2012 and processed at UNIFAG (Integrated Pharmacology and Gastroenterology Unit of the University of São Francisco). After obtaining the samples through biopsies and polymerase chain reaction (PCR), it was possible to determine the presence of allelic variants of the respective genes.

Keywords: Helicobacter pylori. **cagA**. **vacA**. Gastric cancer.

INTRODUCTION

Helicobacter pylori is a gram-negative, microaerophilic, curved spiral bacillus with five to six flagella, which give it great mobility. It can be found in two forms, the spiral form which is around 3-5 pm **long and approximately** 0.5 **pm in** diameter, and the coccoid form. Studies show that this bacterium, in its spiral form, is more active and resistant, while in its coccoid form its viability is questioned (KUSTERS et al., 1997).

There is a protrusion at the end of each flagellum, making it possible for them to move through the thick layer of mucus that covers the stomach lining. Other bacteria with normal flagella are unable to move with such ease, becoming trapped in the mucus. **H. pylori** is able to survive the very acidic conditions of the stomach by producing ammonia from urea. The function of ammonia is to neutralise the gastric acidity around the **Helicobacter** cell, allowing the bacterium to survive and reproduce, colonising and multiplying above the epithelial cell layer that makes up the stomach mucosa (MONTECCUCO, C.; RAPPUOLI, R. 2001).

The first researcher to detect the presence of this microorganism was W. Jaworski, in 1886, analysing human samples taken from stomach washings. The work carried out by this researcher can be found in his book **Handbook of Gastric Diseases,** which suggests for the first time a relationship between the microorganism and gastric diseases (KONTUREK, 2003). Researchers Barry Marshall and J. Warren also cultivated the bacterium for the first time in 1982, previously called **Campylobacter like organisms,** later called **Campylobacter pyloridis** and later, due to grammatical corrections, **Campylobacter pylori**. The name of the bacterium was updated after several taxonomic studies because its biochemical and genetic characteristics showed it to be different from species of the **Campylobacter** genus and very close to the **Helicobacter felis** and **Helicobacter mustelae** species, found in cats and ferrets respectively (HEATLEY, 1995).

In 1994, **H. pylori** was classified as a type I carcinogen by the IARC (**International Agency for Research on Cancer**). However, the role of the bacterium in the genesis of gastric cancer is still controversial, because although there is a high prevalence of infection, less than 1% of infected individuals develop the disease. In 1996 all the material studied by W. Jaworski was discovered and to this day he is considered the forerunner of gastric microbiology. In 1997, the first **Helicobacter pylori** genome was published by Tomp et al. in which his team sequenced strain 26695, isolated from a patient with gastritis. In the United States, they completed the genome of the J99 strain from a patient with duodenal ulcer (HANCOCK et al., 1998).

Studies have shown that the bacterium **Helicobacter pylori** is directly involved in gastrointestinal tissue pathologies. Among the different types of bacteria, the virulence factors **cagA** and **vacA** are considered the most aggressive when in contact with the human organism. According to studies, the bacterium has a high survival rate when in contact with gastric juice, thus demonstrating the great difficulty in eradicating it from the body. Another important point to note is that although **H. pylori** infection affects more than half of human beings, the percentage of carcinomas that develop is low, thus concluding that the environmental factor and the individual's quality of life can enhance or inhibit the progressive action of the bacteria.

In view of these factors, it is necessary to study the main characteristics of **Helicobacter pylori**, especially its most virulent strains, so that procedures and drugs can be developed that are effective both in treating those infected and in preventing possible infections.

The main aim of this study was to analyse the possible association between the **cagA** and **vacA** virulence factors of the **Helicobacter pylori** bacterium and the development of gastrointestinal pathologies. To this end, a survey of biopsies taken through endoscopic procedures was proposed, in which the aforementioned virulence factors and the respective pathologies of each patient were analysed.

According to the analyses, all the samples tested positive for the bacterium and of these results the virulence factors **cagA** and **vacA** had considerable expression, corroborating current literature. In all cases there was a gastric pathological diagnosis, which could characterise the influence of the bacteria in the development of certain diseases.

CHAPTER 1

PREVALENCE AND ASSOCIATION OF HELICOBACTER PYLORI WITH DIGESTIVE TRACT DISEASES

1.1 Epidemiology

Helicobacter pylori is present in 95 per cent of patients with duodenal ulcers and in 70 per cent of those with gastric ulcers. In developed countries, it is not common to find infected children, but the infection rate increases by around 1% per year of age over 20, eventually reaching an average of 20 to 30% among adults in the United States. In underdeveloped countries, it is commonly found in children and reaches levels of 80 per cent in the adult population. In Latin America, stomach cancer rates are the highest in the world, as is the rate of infection with the bacterium. Considered to be one of the most frequent chronic bacterial infections in humans (GRAHAM, 2000), it is believed that 50% of the world's population is infected with the bacterium (figure 1) (ROBERTSON et al., 2003) and that in Brazil the rates can reach 80% (OLIVEIRA et al. 1999; RODRIGUES et al., 2004).

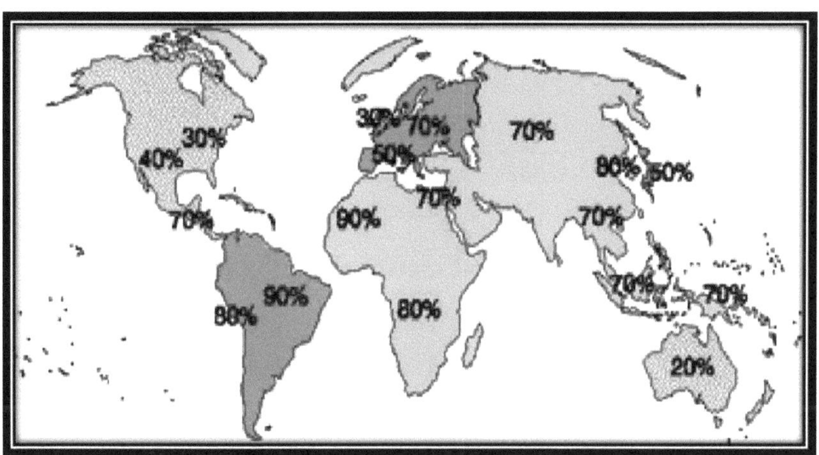

Figure 1. Worldwide prevalence of H. pylori infection.
Source: Helicobacter **foundations (http://www.helico.com/h_epidemiology.html)**

Another interesting fact is that **H. pylori** infection is one of the most common in humans (Atherton; Blaser, 2009). As a result, proven genomic studies point to the great genetic diversity of this bacterium (Dorer et al., 2010). Although the rate of infection in developed countries, such as the United States, is relatively low, the

Hispanics living in this country have an infection rate of up to 75 per cent. When it comes to the East, Japan has lower rates of stomach cancer and **H. pylori** infection at the same time. Perhaps having chronic inflammation of the stomach over a long period of time - which occurs when a person suffers from **H.** *pylori* infection early in life - predisposes them to stomach cancer. As a result, one line of researchers is reluctant to consider **H. pylori** as a definitive agent of stomach cancer and believe that it may only be a co-factor, as most infected people do not develop the disease. However, the most common type of stomach cancer (gastrointestinal) has an 89% correlation with **Helicobacter pylori** infection (INCA, 2003).

This microorganism can be acquired in childhood around the age of 10 and it is estimated that more than 50% of the world's children are carriers (POUNDER, 1995). In developed countries, where sanitary conditions are better, the infection is infrequent in childhood, but the incidence increases with advancing age at a rate of 0.5 to 1.0 per cent per year and around 50 per cent of the population over 60 has the infection. In developing countries, infection occurs at an early age and with higher frequencies in childhood, reaching 80% or more of adults (figure 2) (BROOKS, 2000; TRABULSI et al.,2000).

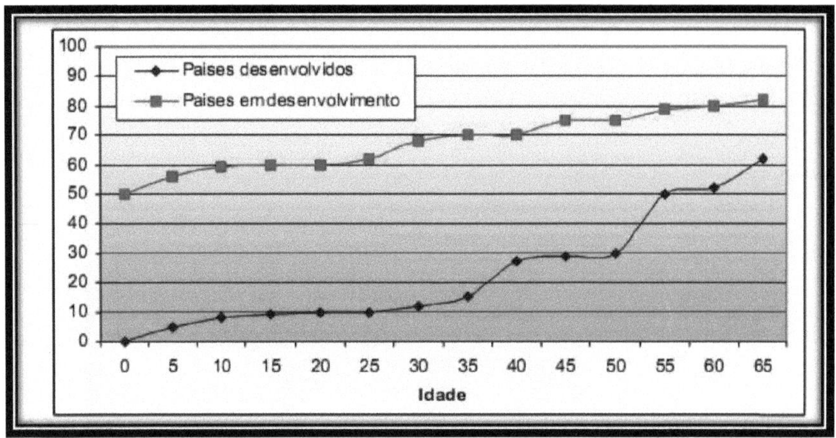

Figure 2: H. pylori infection rate.
Source: *Helicobacter* **foundations** (http://www.helico.eom/h epidemiology.html).

In Brazil, there are few studies on the prevalence of the bacterium. There are data from the state of Mato Grosso, where the rate of **H. pylori** infection among children and adolescents is high.
in adolescents reaches 77.5 per cent and 84.7 per cent in adults (SOUTO et al., 1998). In the state of Minas Gerais, more specifically in Belo Horizonte, a study showed a prevalence of the bacteria of 34.1% in children and 81.7% in adults (OLIVEIRA et al., 1999). Also in the northern region of Brazil, in the state of Ceará, the increase in **Helicobacter pylori** infection rates has grown substantially in low-income populations, reaching 35% in children up to 12 years of age and from 12 to 14 years of age the rate reaches 74.5% of those infected (ROGRIGUES et al., 2004).

1.2 Helicobacter pylori **and gastric cancer**

Gastric cancer, commonly known as stomach cancer, is predominantly characterised by three

8

histological types: Adenocarcinoma (responsible for 95% of tumours), Lymphoma (diagnosed in a certain 3% of cases) and Leiomyosarcoma (which begins in the tissues that give rise to muscles and bones). The highest incidence rate is among men, around the age of 70, and approximately 65 per cent of patients diagnosed with stomach cancer are over the age of 50 (INCA, 2012).

Some pre-existing diseases can be associated with gastric cancer, such as pernicious anaemia, precancerous lesions (such as atrophic gastritis and intestinal metaplasia), and infections with the bacterium **Helicobacter pylori**. The widespread contamination of food and drinking water by **H.** *pylori is* considered to be the second most frequent, surpassed only by caries bacteria. Estimates suggest that the bacterium inhabits the stomach of around 70 per cent of the population in Brazil, but only genetically predisposed individuals, i.e. those born with a receptor in the stomach capable of hosting the bacterium, are affected. An extremely important factor in the development of gastric cancer, together with **H. pylori** infection, is the quality of the individual's diet (low intake of fruit and vegetables, excess salt, nitrites and nitrates) and lifestyle (KELLEY; DUGGAN, 2003).

Studies show that smoking, excessive alcohol consumption and a family history of cancer can increase the risk of the disease by up to 16 times more than uninfected individuals with no family history, making the bacterium a co-factor in the development of gastric diseases. This means that not everyone infected with the bacteria will develop gastric cancer, but the vast majority of patients diagnosed with the disease are positive for **Helicobacter pylori** (SEPÚLVEDA, 2001).

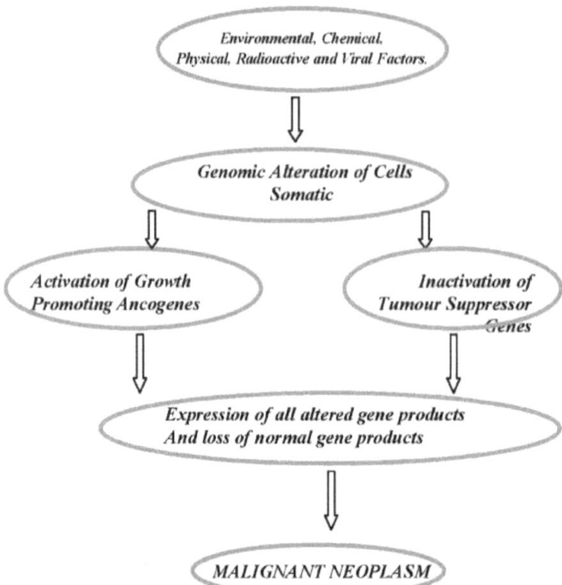

Fig. 3 illustrates the development of a malignant neoplasm, while Fig. 4 shows a stomach with cancer.

Figura 3: **Description of the development of malignant neoplasia Source: TOMANINI, R. Bioquímica do câncer, 2011.**

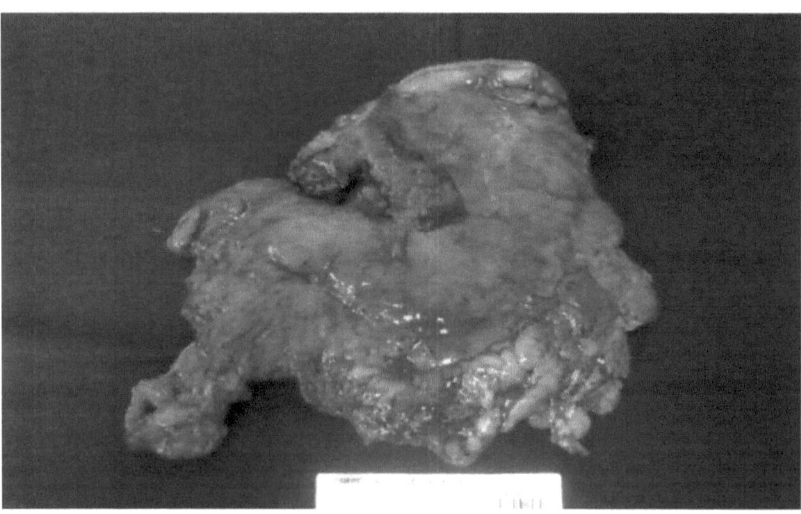

Figura 4: **Photo of stomach with advanced cancer, removed during gastrectomy. Source: Hospital São Francisco surgical centre, 2011.**

1.3 Mild, moderate and severe antral gastritis

The Sydney System for the classification of gastritis emphasised the importance of combining topographical, morphological and aetiological information into a scheme that would help generate reproducible and clinically useful diagnoses. To re-evaluate the Sydney system 4 years after its introduction, a group of gastrointestinal pathologists from various parts of the world met in Houston, Texas, in September 1994. The objectives of the seminar were to establish a consensus terminology for gastritis, identify, define and try to resolve some of the problems associated with the Sydney system (DIXON at al., 1996).

Overall, the principles and classification of the system were only slightly modified and the classification was adjusted by providing a visual analogue scale. The terminology of the final classification was improved to emphasise the distinction between atrophic and nonatrophic stomach, where the names used for each entity were selected because they are generally acceptable to both pathologists and gastroenterologists. In addition to the main categories, atrophic and nonatrophic gastritis, the special or distinctive forms are described and their respective diagnostic criteria are provided. The article included practical guidelines for optical biopsy sampling of the stomach, for using the visual analogue scales for grading histopathological features and for formulating a standardised complete diagnosis. A glossary of terms related to gastritis was used for reference (DIXON at al., 1996). According to the Sidney System, gastritis can be classified endoscopically (Table 1).

Table 1: Sidney system for classifying gastritis

TOPOGRAPHY	CATEGORY	DEGREE OF INTENSITY
Pangastritis	Enchanting	Lightweight
Gastritis of the antrum	Flat Erosive	Moderate
Gastritis of the body	Erosive High	Severe
	Atrophic	
	Haemorrhagic	
	Reflux	
	Pleats	
	Mucosa	
	Hyperplastic	

Descriptive terms:
- Oedema
- Enanthema
- Friability
- Exudate
- Flat erosion
- High erosion
- Nodosity
- Mucosal Fold Hyperplasia
- Mucosal fold atrophy
- Vascular pattern visibility
- Areas of Intramural Haemorrhage

Source: Gastrocentre - UNICAMP

There is no consensus on the definition of gastritis, as the term is used by endoscopists when diagnosing it visually, by radiologists in terms of changes in the silhouette of the mucosa seen by X-ray, by pathologists who define it through microscopic analyses and by doctors, who have no objective method, through diagnoses of family history, smoking, alcoholism and characteristic symptoms. The accuracy of the diagnosis of gastritis is confirmed by microscopic examination, which analyses a biopsy taken from the patient during an endoscopic examination. Once an acute **Helicobacter pylori** infection has occurred, almost all patients will develop chronic gastritis, most of which will remain asymptomatic throughout their lives. Less than 20 per cent of patients with chronic infection will develop peptic ulcer disease and a small percentage will develop gastric adenocarcinoma and gastric lymphoma (PETERSON, 2002).

1.4 Helicobacter pylori **and pangastritis**

Pangastritis is directly associated with general inflammation of the stomach mucosa, as "pan" means whole, gastritis means inflammation of the mucosa (inner layer) of the stomach, i.e. inflammation affecting the entire mucosa of the stomach with a reddish appearance and swelling of medium intensity. Some significant inflammatory lesions of the gastric body are responsible for a decrease in acid secretion after eradication

Helicobacter pylori can reduce the risk of developing gastroesophageal reflux disease (GERD) (KORWIN, 2001).

1.5 **Reflux and** Helicobacter pylori

Gastro-oesophageal reflux is a chronic disease due to the retrograde flow of part of the gastroduodenal contents into the oesophagus, which may or may not be associated with tissue damage. Data indicates that Brazil has approximately 12 per cent of the population suffering from reflux disease (MORAES FILHO et al., 2002). Evidence suggests that there is an increase in the incidence of gastroesophageal reflux (GER) after eradication of **H. pylori** from the gastric mucosa (SOUZA; LIMA, 2009). It is speculated that gastric infection with the bacterium **may exert some "protective" effect against the development of oesophagitis and** Barrett's **oesophagus** (lesions recognised as precancerous), but the association between GER after **Helicobacter pylori** eradication and the increased incidence of oesophageal adenocarcinoma **has** not been established (KODAIRA, 2002).

1.6 Helicobacter pylori **and peptic ulcer disease**

Experimental and indisputable evidence indicates that **Helicobacter pylori** is an important agent in the genesis of ulcers, both because of the inflammation of the mucosa due to the presence of the bacteria and because of the alteration of the mechanisms that regulate acid production. Around 90 to 95 per cent of patients with ulcers are infected with the bacteria. Studies show that once **H. pylori** has been eradicated from peptic ulcers, the patient can be said to be cured of the ulcer, except for other less frequent factors such as the use of non-hormonal anti-inflammatory drugs and acetylsalicylic acid (EISIG, J.N.; CARVALHAES, A. 2006).

1.7 Methodologies and diagnoses

Among the methods of analysis for diagnosing **Helicobacter pylori** are the rapid urease test and the breath test. The breath test consists of a simplified 14C-urea rapid breath test for diagnosis. The fasting patient undergoes the initial assessment for **H. pylori** by drinking 5 microCi of 14C-urea in 20 ml of water. Breath is collected at 30 min intervals. The samples are counted in a beta-counter, and the results are expressed as counts per minute (cpm) (MARSHALL et al., 1991).

With regard to the urease test, to which the samples from the patients in this project were submitted, it is known that the urease gel is made up of urea 20.0g/L; disodium phosphate 1.8g/L; phenol red 0.012g/L; bacteriological agar 3.5g/L and purified water 1000mL (figure 8, page 23). Urease is an enzyme produced by many species of microorganisms that hydrolyse urea to form ammonia and carbon dioxide. The ammonia reacts in solution to form ammonium carbonate, resulting in alkalinisation of the environment. **Helicobacter pylori** produces large quantities of urease, and this pre-formed enzyme can be detected in biopsy fragments from the body and gastric antrum using a rapid test for urea hydrolysis. The result is interpreted by looking at the colour of the gel, where a positive test is the development of an alkaline reaction (strong red or pink colour) and a negative test is the absence of a change in the colour of the gel (yellow or light orange) (BRANSON,D. 1973).

1.8 Forms of treatment

Treatment with antimicrobials for **H. pylori** infection results in the healing of ulcers and can prevent their recurrence. For this reason, there is interest in obtaining results on the possible mechanisms by which this microorganism affects the balance of mucosal defences. Studies indicate that **H. pylori** secretes urease, generating free ammonia, and protease, which fragments glycoproteins in gastric mucus (SMOOT et al, 1990; MITANI-EHARA, 1994; BRZOZOWSKI et al, 1996; MITANI-EHARA et al, 1997; COTRAN et al, 2000; MOBLEY et al, 2001; MURRAY et al, 2007).

Microorganisms also synthesise phospholipases, which damage the superficial epithelial cells and lead to the production and release of eicosanoids (chemical mediators of the inflammatory process). There is also the possibility that neutrophils attracted to **H. pylori** release myeloperoxidase, which produces hypochlorous acid, in turn producing monochloramine in the presence of ammonia. Both hypochlorous acid and monochloramine can damage and destroy mammalian cells. Both mucosal epithelial cells and the endothelial cells of the lamina propria are important targets of the destructive actions of colonisation by the bacteria, where thrombotic occlusion of the superficial capillaries is also caused by a bacterial platelet activating factor. In addition to the synthesis of enzymes by **H. pylori**, other antigens recruit inflammatory cells to the mucosa, which is chronically inflamed, making it more susceptible to gastric acid damage. In this way, it is believed that damage to the mucosa allows tissue nutrients to leak to the surface of the microenvironment, thus sustaining the survival of the bacilli (SMOOT et al, 1990; MITANI-EHARA, 1994; BRZOZOWSKI et al, 1996; MITANI-EHARA et al, 1997; COTRAN et al, 2000; MOBLEY et al, 2001; MURRAY et al, 2007).

Since the relationship between Helicobacter pylori and various pathologies, especially gastrointestinal ones, has been recognised, the indications for its treatment have been expanded (MARSHALL & WARREN, 1984; CURRENT EUROPEAN, 1997; MALFERTHEINER et al, 2002).

Currently, in addition to gastroduodenal ulcers, eradication of the microorganism is recommended in other conditions, such as "mucosa-associated lymphoid tissue" lymphoma (gastric MALT), first-degree relatives of gastric cancer patients and iron deficiency anaemia of obscure etiology, among others (MALFERTHEINER et al., 2007). There is still no 100% effective antimicrobial treatment for eradicating H. pylori infections. What does exist are treatment protocols that simultaneously use antimicrobials with a spectrum of action for the bacteria, proton pump inhibitors (PPIs) and urease inhibitors. However, these treatment protocols do not have the same result for all cases (BRUNTON et al, 2006; MURRAY et al, 2007; SIQUEIRA et al, 2007).

According to studies, microbial sensitivity varies according to geographical location, race and previous use of these drugs, thus jeopardising the effectiveness of a treatment regimen in a community and making the results unreliable. The ideal approach is to develop treatment based on sensitivity studies or at least prior knowledge of the rate of microbial resistance. Among the factors that contribute to the anti-H. pylori therapeutic problem are those related to the formulated regimen, antimicrobial resistance (especially related to clarithromycin and imidazoles), lack of adherence to treatment, geographical factors and the dose of proton pump inhibitor used. Therapy with just one antimicrobial usually doesn't bring good results; the dual antimicrobial regimens initially tested in paediatrics lasted at least four weeks and included the combination of bismuth with amoxicillin (or ampicillin) or amoxicillin with tinidazole. However, in the hospital environment, where there is a high percentage of resistant strains, it is very likely that a dual treatment that includes these drugs will not be effective (BRUNTON et al, 2006; MURRAY et al, 2007; SIQUEIRA et al, 2007).

Since 1989, a combination of amoxicillin (50mg/kg/day), metronidazole (20 to 30 mg/kg/day) and furazolidone (6 to 8 mg/kg/day) has been used in 3 daily doses for 7 days, although the percentage of side effects is significant. One of the best approaches to therapeutic treatment is to use an antisecretory agent, usually a proton pump inhibitor (PPI), combined with two antibacterials for a period of 7 to 14 days. The association of PPIs with antimicrobials acts effectively, potentiating the bactericidal effects of the regimen (LIND et al., 1999). As a basic criterion, the first regimen against **Helicobacter pylori** infection is the one with the greatest chance of eradicating the infection. However, the ineffectiveness of these treatments occurs in 15 to 20 per cent of cases, determining the need for a new intervention (CHEY & WONG, 2007). As a second line of treatment for the infection, it is recommended to use a therapeutic procedure with four lines of drugs, for at least 7 days, combining a proton pump inhibitor (PPI), colloidal bismuth subcitrate/subsalicylate, metronidazole and tetracycline.

Currently, the combination of a proton pump inhibitor (omeprazole 20 mg, lansoprazole 30 mg, pantoprazole 40 mg, rabeprazole 20 mg) with amoxicillin 1000 mg and clarithromycin 500 mg is used. This triple treatment has been more effective than the four-procedure approach, which suggests the following therapeutic treatment with a proton pump inhibitor (2 times a day) + 500mg of tetracycline (4 times a day) + bismuth subsalicylate or subcitrate (4 times a day) + 500mg of metronidazole (3 times a day). Patients who remain infected despite treatment, including macrolides and nitroimidazoles, often develop resistance to at least one of these antibiotics (BRUNTON et al., 2006).

It is important to emphasise that the successful eradication of **Helicobacter pylori** does not depend exclusively on the susceptibility of the bacterial strain to the antimicrobial used in treatment, although this is one of the main factors, especially if we consider patients infected with antimicrobial-sensitive strains (SIQUEIRA et al, 2007).

It is worth emphasising that effective antimicrobial therapy for this infection presents numerous challenges. There is a consensus that symptomatic individuals should be treated for the infection so that the most serious complications of the disease can be prevented. Treatment of symptomatic patients, in the absence of a definitive diagnostic result in therapy, is unnecessary for microbe-negative patients with disorders such as non-ulcer dyspepsia, increasing the risk of antimicrobial resistance (KRAKOWKA et al., 1998).

The next chapter will look at the main virulence factors of the bacteria and their influence on the development of diseases of the digestive tract.

CHAPTER 2

VIRULENCE FACTORS

A few decades ago, around forty epidemiologically unrelated Helicobacter **pylori** strains were identified by analysing their genome and plasmid diversity. These genomic variations in strains may be responsible for encoding different virulence factors, in a position to develop different types of pathologies in the host through the release of toxins. The increase in diversity between strains is due to the ability to obtain exogenous DNA, contributing to genomic diversity, such as the **cag** pathogenicity island (**cag-PAI**), active in many **H. pylori** strains (ATHERTON et al., 1995; VAN DOORN et al., 2000).

Among the most studied virulence factors is **cagA**, which is considered positive or negative in the samples (YAMAOKA, 2010). This virulence factor is associated with an increase in gastric tissue inflammation and a higher risk of developing cancer (PEEK; CRABTREE, 2006; ERNST et al., 2006).

The **vacA** virulence factor is the second most studied (YAMAOKA, 2010) and its characteristic is to encode a secretory protein that is able to induce cell vacuolisation by increasing permeability to anions and urea (TOMBOLA et al., 2001).

The bacterium also has a mechanism of resistance to some antimicrobials, **ureC** and **iceA,** which are point mutations associated with the diversity detected in some genes (VAN DOORN et al., 2000).

Environmental influences together with a set of virulence factors trigger a complex network that regulates diseases of the digestive tract, which can lead to the development of inflammatory processes and gastric carciogenesis (ISOMOTO et al., 2010). Studies have established that cancer is a pathology that involves several factors and that the oxidative stress developed by inflammation of the gastric wall, the production of oncogenic proteins, epigenetic systems, exposure to the environment and genetic susceptibility can together cause carcinogenic development. Infection with **H.** *pylori and* the consequent chronic inflammation caused by the bacterium activate various oncogenesis pathways that promote the development of stomach cancer (DING et al., 2010).

2.1 vacA gene and its subtypes

Encoded by the **vacA** gene, the production of vacuolating cytotoxin is an important virulence factor of **H. pylori** and induces cytoplasmic vacuolisation in eukaryotic cells (ATHERTON et al., 1995; BLASER, 1992; COVER et al., 1994; MONTECUCCO et al., 1999). This cytotoxin also acts on the immune system, interfering with the action of antigens and inhibiting the production of antineoplastic T defence cells by different mechanisms (BETTEN et al., 2001).

Epidemiological analyses show in animal experiments that the cytotoxin **vacA** is the major virulence factor related to damage to the gastric mucosa. Because it has allelic heterogeneity and is secreted by all strains of the bacterium, **vacA** may be a determining factor in the variations in clinical manifestations among patients infected with **H. pylori**. **VacA** is secreted into the extra-cellular space and is also partially retained on the bacterial cell surface. At neutral pH, it is grouped into a large, water-soluble oligomeric complex made up of 12 or 14 identical monomers (LUPETTI et al., 1996; COVER et al., 1997).

The protein induces vacuolisation, changes in endolysosomal function, the appearance of pores in the plasma membrane, permeabilisation of the epithelial layer, as well as cell apoptosis and affects different regions of the cell such as epithelial junctions, the cytoskeleton, mitochondria and endocytic vesicles (Figure 5) (PAPINI et al., 2001).

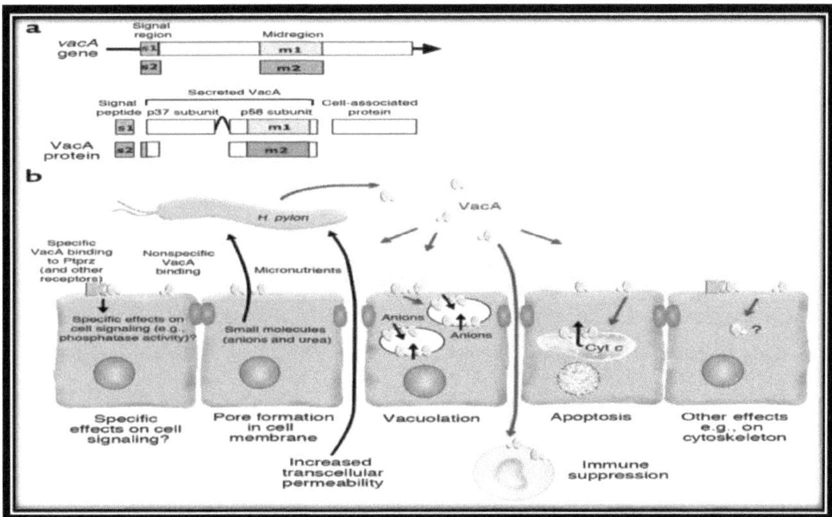

Figure 5: Action of H. pylori vacA on the gastric mucosa Source: BLASER, M.J.; ATHERTON, J.C. , 2004.

The **vacA** gene has two variable regions, one of which is approximately 50pb, called **s, located in the 5' terminal region, coding for the signal peptide** and in which two subtypes **s1** and **s2** can be identified. In another region with great heterogeneity, located in the middle part of the gene, we find the site called **m** with approximately 700pb, where two subtypes **m1** and **m2** can also be found. What determines the production and specificity of cytotoxin activity, associated with the pathogenicity of the bacteria, is the mosaic combination of these distinct allelic types (s and m) (ATHERTON et al., 1995; GUNN et al., 1998; VAN DOORN et al., 1998; WANG et al., 1998). All strains have one of the sequence types and one of the two median region types, and may have combinations of s1m1, s1m2, s2m2 and, the least common, s2m1. Analyses and studies show that in the s1m1 genotype, vacuolising activity is very high compared to the s1m2 genotype. In the s2m2 genotype, however, this activity is absent. Because of this, the presence of the s1m1 mosaic is often associated with the development of peptic ulcers and gastric carcinoma (ESKANDAR et al., 2006). The proteins produced from **vacA** m1 and **vacA** m2 can react with different receptors on human gastric epithelial cells, eliciting individual and particular physiological responses (PAGLIACCIA et al., 1998).

There is no common consensus in the literature on the relationship between **vacA** s1m1 and the development of peptic ulcers. On the other hand, other authors have linked **vacA** mosaics to the presence of duodenal ulcers (ARENTS et al., 2001).

2.2 Gene cagA

The **cagA (cytotoxin antigen associated)** strain-specific gene, the first to be identified, is directly related to the progression of gastrointestinal diseases, as it is one of the most virulent forms and induces increased levels of some cytokines, such as IL-ip and IL-8, with an increased risk of developing gastritis and gastric cancer (STOICOV et al., 2004), showing that patients infected with this strain have a 2.87 times greater risk of developing gastric cancer. The cagA gene has the characteristic of being a marker of the **cag** pathogenicity island (**cag-PAI**). It is approximately 40 Kb long and encodes around 31 genes. It is found in around 50 to 60 per cent of the strains that appear in the West and in 90 per cent of those that occur in Japan (NAITO and YOSHIKAWA, 2002).

A characteristic of the **cag-PAI** island is that it contains homologous genes and genes from other bacteria that encode components of the secretion system, forming a syringe-like structure with the function of injecting the **cagA** protein and possibly other unknown virulence factors into the host's epithelial cells (HATAKEYAMA, 2004) (Figure 6). This protein, considered to be oncogenic, is phosphorylated by members of the SRC family of kinases, where once phosphorylated it binds to the SHP2 protein (SRC - homologue 2) which acts in the translation of mitogenic signals via RAS-MAPK, which can lead to cell multiplication, apoptosis and rearrangement of the cell's cytoskeleton, altering its morphology, making it similar to a **hummingbird**, described as the **hummingbird phenotype** (HATEKEYAMA, 2004).

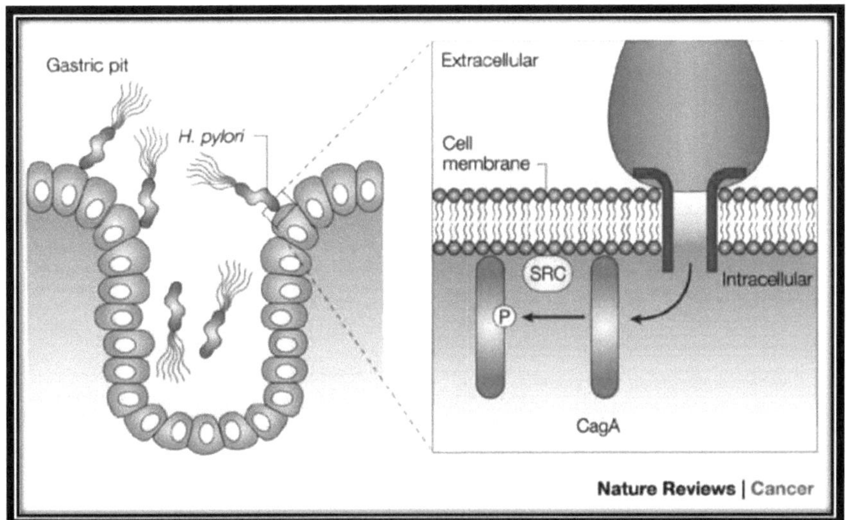

Figure 6: Interaction between H. pylori cagA **and gastric epithelial cells Source: HATAKEYYAMA, M. 2004.**

2.3 Helicobacter pylori **colonisation**

Factors that characterise the bacterium in its morphology are extremely important for its colonisation and survival in the digestive tract. The presence of unipolar flagella and their spiral shape allow the bacteria to move from the lumen of the stomach, where the pH is low, to an area where the pH level is close to neutral, thus providing favourable conditions for their development and growth in the form of colonies. These flagella, of which there may be five or six, are made up of flagellin subunits (a protein structure) and covered in a double phosphorolipid layer whose specific function is to protect the bacteria from gastric acidity (EATON et al., 1989; KOSTRZYNSKA et al., 1991; LEYING et al., 1992; SUERBAUM et al., 1993). This characteristic of a flagellated microorganism allows **Helicobacter pylori** sufficient mobility and resistance to penetrate the mucus layer of the stomach wall and establish contact with epithelial cells (JENKS, 2000) (Figure 7).

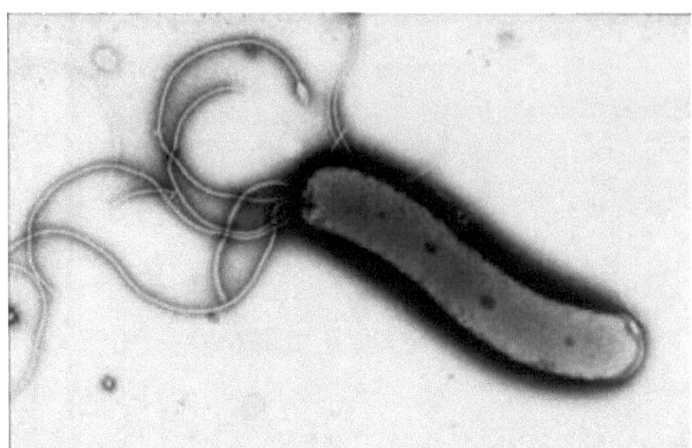

Figure 7: Helicobacter pylori
Source: http://bioweb.uwlax.edu

A biochemical characteristic of **Helicobacter pylori** is the high production of urease, which is extremely important in its colonisation, where it favours adaptation to the environment, as this protein has a high molecular weight (500 to 600KDa) and promotes the hydrolysis of urea, present in gastric juice, resulting in ammonia (WEEKS, 2001).

With this product, the bacteria become internally neutral in their pH, due to the influence of H+ ion receptors, thus allowing greater resistance to gastric acidity. This transformation of urea into ammonia also serves as a source of nitrogen by synthesising proteins, which are indispensable for the bacteria's adherence. Together with ammonia, urease promotes the instability of the mucus layer, leading to the formation of lesions in the lining cells and consequently exposing the inner wall of the stomach to contact with gastric juice. The activation of monocytes, neutrophils and the immune system by urease may be directly related to the formation of specific inflammatory lesions (EATON, 1995).

2.4 Transmission media

Numerous studies have pointed to some possible risk factors associated with gastric cancer, such as ethnicity, age, gender, smoking, socio-economic **status**, excessive consumption of salt, nitrates, nitrites and alcohol (KELLEY; DUGGAN, 2003). However, **Helicobacter pylori** is one of the main risk factors for the development of gastric cancer (CHOLI-PAPADOPOULOU et al., 2011).

The exact transmission characteristics of **Helicobacter pylori** are still unknown, and the only fact accepted worldwide is that the bacterium can only reach the gastric mucosa through the mouth, as it is a non-invasive microorganism. The microorganism in question is fragile under laboratory conditions, suggesting that its action is limited outside the host. The high prevalence rates in individuals living in crowded and unsanitary conditions suggest that person-to-person transmission is an important mechanism of infection. However, it is not yet possible to determine exactly whether the main route of transmission is oral-oral or faecal-oral. It is likely that both act simultaneously at population levels (KODAIRA, 2002).

Transmission via the oral-oral route can be detected in the oral cavity by means of culture or PCR of materials such as saliva and dental plaque. The oral cavity has been suggested as a reservoir for **Helicobacter pylori** infection and reinfection, as regurgitation of gastric juice can contaminate the oral cavity, predisposing it to colonisation by this bacterium for an undetermined period of time. Furthermore, it has been observed that treating the infection systemically does not eradicate the agent in dental plaque, allowing the mouth cavity to act as a permanent reservoir for the bacteria (KODAIRA, 2002).

In the literature, several studies suggest that the oral-oral route plays an important role in the transmission of the bacteria. Goodmam et al., 2011, in Colombia, detected a higher frequency of infection among individuals who drink from glasses previously used by other people and not washed. Similar transmission mechanisms occur in China, where people have a habit of eating from the same container, favouring the transmission of **Helicobacter pylori** through contaminated chopsticks. The same authors also mention that the fact that mothers blow on food before offering it to their children also plays a significant role in the transmission of the bacteria to children. Theoretically, **Helicobacter pylori** can also be transmitted via the oral-oral route between husband and wife through saliva contaminated by gastric juice. Some authors have described cases of patients reinfected with strains identical to those of their asymptomatic wives, suggesting that infection and reinfection can occur through person-to-person dissemination.

The bacterium has a high degree of genotypic diversity, and most individuals practically carry strains with a unique pattern. The concordance of molecular types found between members of the same family indicates that transmission occurs frequently between these individuals. However, the possibility cannot be ruled out that these individuals may have acquired the disease through exposure to the common source of infection (KODAIRA, 2002).

In faecal-oral transmission, the age-related hepatitis A prevalence curve is considered a marker for the transmission of infectious agents through faecal-oral contamination. There is a strong parallel between the profiles of the **H. pylori** and hepatitis A curves in most of the populations studied, particularly those in developing countries. Following studies by Thomas et al. in 2000, other authors were able to isolate the agent in faeces from patients who did not have achlorhydria or diarrhoea. In England, they cultured **Helicobacter pylori** in the faeces of 12 (48%) of 25 colonised individuals. The high success rate compared to previous studies was attributed to the method used. The authors used the supernatant obtained after centrifuging the faeces, rather than culturing all the faecal material. Despite the fact that the bacteria can be eliminated through faeces, the exact mechanism of transmission of the agent by this route is unknown, as is the real epidemiological importance of this process (KODAIRA, 2002).

In relation to faecal-oral transmission at a population level, the spread of infectious diseases through water is based on its contamination by faeces. Studies have detected the presence of **Helicobacter pylori** by PCR in treated sewage water in the city of Lima, Peru, and studies carried out in Colombia have also identified **H. pylori** in drinking water by PCR. Despite this evidence, more in-depth studies are still needed to prove the viability of this organism in nature (KODAIRA, 2002).

CHAPTER 3

METHODS AND MATERIALS

We analysed 45 samples from patients of both sexes, aged between 18 and 60, from the Integrated Pharmacology and Gastroenterology Unit of the University of São Francisco (UNIFAG-USF) under the authorisation and responsibility of Prof. Dr. Marcelo Lima Ribeiro. Patients with previous gastric surgery, who had been treated or were being treated with antibiotics to eradicate the microorganism, who had been taking anti-inflammatory drugs and proton pump inhibitors in the last 3 months, drug users, smokers, as well as those who refused to take part were excluded. Those who agreed to take part in the study completed a consent form. A biopsy of the antrum region was carried out using an endoscopic procedure to diagnose the presence of the bacteria, initially using a urease test. Although there are negative results for the bacteria in this test, it is known that false negative results are common due to the low virulence of the bacteria in certain cases. For this reason, all the biopsies were subjected to DNA extraction according to the phenolchloroform protocol (Fox et al. 1994) (appendix 01) and after this procedure, they were subjected to PCR (**Polymerase Chain Reaction**) (appendix II) for positive confirmation of the bacteria and then the characterisation, also by PCR, of the **cagA** and **vacA** genotypes (BRANSON,D. 1973).

The medical report with the appropriate diagnoses was provided by the doctor in charge.

3.1 Histology

A biopsy was extracted from the patients' gastric antrum and immediately placed in a urease substance (Figure 4) to detect the bacteria. All the samples, positive and negative for urease, were taken to the laboratory for initial DNA extraction and later analysis by PCR. In the PCR stage, positive and negative samples for **H. pylori** were diagnosed, and negative samples were discarded. The positive samples were subjected to specific PCR for the **cagA** and **vacA/m1-m2**, s1-s2 genes.

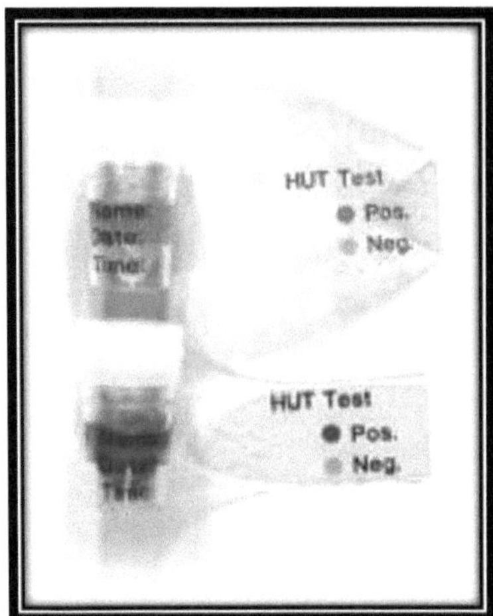

Figure 8: Urease kit for Helicobacter pylori **testing Source: medicaldiagnostics.net.au**

3.2 Patients sampled

Samples were analysed from 45 **Helicobacter pylori** positive patients of both sexes, aged between 18 and 60, who had gastrointestinal problems. The endoscopic procedure was used to assess the presence of oesophagitis, gastritis, peptic ulcers and lesions that could progress to carcinomas. In cases where malignancy was suspected, gastric biopsies were taken and reports issued by the doctor in charge. The reports were sent to the UNIFAG laboratory for comparison and analysis with the results obtained from the biopsies.

3.3 DNA extraction

One biopsy from each patient was used for genomic DNA extraction, following the phenolchloroform protocol (Fox et al. 1994). In summary, the collected biopsies were **immersed in 300|il of buffer solution (EDTA 1 mM pH8.0; Tris-HCL** 50mM pH 8.0; Tween 0.5%), added 15mg/ml of proteinase K and incubated at a temperature of 37°C for a minimum of 5 hours. Afterwards, the DNA was purified and precipitated with 3M sodium acetate by centrifugation with chloroform phenol. Quantitative DNA results were then obtained using Nanodrop equipment with a computer, **giving results in ng/pl.**

3.4 PCR (Polymerase Chain Reaction) and Electrophoresis Analysis

The PCR technique is based on the concept of enzymatic amplification of a specific DNA site by extending two oligonucleotides, called **"primers", which hybridise with the complementary strands of a target sequence.** The template DNA is denatured using 35 cylclos at specific temperatures (Figure 9), enabling the **primers** to hybridise to **their complementary sequences and consequently the extension of the primers hybridised** by the DNA polymerase. **The reagents and** their respective concentrations are shown in Annex II. The end result is the exponential accumulation of **a specific fragment, defined by the 5' end of the primers. After obtaining the PCR product, the samples were subjected to electrophoresis** (appendix III) using a 0.8% agarose gel and a 100V, 110M.A. vat for a 1-hour run. The gel results were then placed in a photodocumentation system (UV and white light transilluminator attached) with a computer for **photographic recording of the results. The primers, described in Table 2, were developed** for specific DNA fragments, thus making the results obtained after the whole process more reliable.

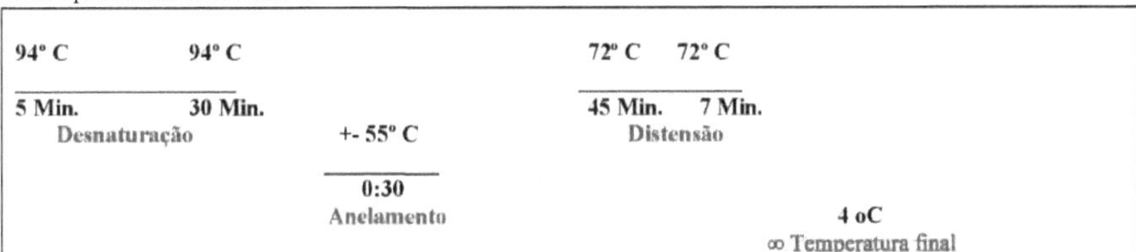

Figure 9: DNA hybridisation cycle (35X) in the thermal cycler (PCR).
Source: TOMANINI, R., 2012.

Table 2: Specific primers **for obtaining results.**

Gene	Primer	Sequence (5 -▶ 3')	PCR (bp)
Shit	D008-F	ATAATGCTAAATTAGACAACTTGAGCGA	297 *
Shit	R008-R	TTAGAATAATCAACAAACATCACGCCAT	297 *
VacA m1	VA3-F	GGTCAAATGCGGTCATGG	290 **
VacA m1	VA3-R	CCATTGGTACCTGTAGAAAC	290 **
VacA m2	VA4-F	GGAGCCCCAGGAAACATTG	352 **
VacA m2	VA4-R	CATAACTAGCGCCTTGCAC	352 **
VacA s1/s2	VA1-F	ATGGAAATACAACAAACACAC	259/286 ***
VacA s1/s2	VA1-R	CTGCTTGAATGCGCCAAAC	259/286 ***

Source: *COVACCI, 1993;** TUMMRU, 1993;*** ATHERTON, 1995.

CHAPTER 4

RESULTS AND DISCUSSION

The great genetic diversity is a striking feature of gastric **Helicobacter pylori**. Studies indicate that half of the human population is infected with the bacterium, and it is a pathogen responsible for developing chronic gastritis, which can lead to peptic ulcers of the stomach and duodenum, gastric atrophy, adenocarcinomas and lymphoid tissues (MALT). This considerable number of diseases may be the result of the great allelic variety and wide genetic variability characteristic of this species, since the combination of the high mutation rate associated with the frequent exchange of genetic material during infection in the stomach results in multiple strains of the bacterium (KENNEMANN et al., 2011).

With this principle in mind, this study analysed the incidence of **H. pylori** and its most frequent genes in patients with pathologies of the digestive tract. We analysed the expression of 05 genes which, alone or in a mosaic, are directly involved in cellular processes such as apoptosis, cell cycle and DNA repair. As has already been shown, there are other factors such as diet quality, smoking and genetic predisposition, which together with the bacterial infection, can cause lesions in the stomach, leading the individual to develop certain pathologies.

Helicobacter pylori has a potent urease activity that participates in gastric colonisation, allowing it to survive in an acidic environment, in this case the human stomach. Urease hydrolyses urea, present in the stomach, into ammonia and CO2 (VOLAND et al., 2003). Ammonia has cytotoxic activity, increasing the permeability of the epithelial cell to protons, and perhaps mediates the local immune response, since it is a potent activator of monocytes in vitro (GOBERT et al., 2002).

The urease test on tissue biopsies is based on the high urease activity of the bacteria. It is highly specific and sensitive and is consistently used for endoscopic diagnosis. A biopsy fragment from the antrum is introduced immediately after collection into a substrate containing urea and a pH indicator (phenol red). Urease hydrolyses the urea into ammonia and carbon dioxide, with a consequent increase in pH and a change in the colour of the medium from yellow to pink. When the colour change occurs within the first 24 hours, the test is considered positive (ORNELLAS et al., 2000).

Although there are various manipulations available on the market, the unbuffered urease test has shown satisfactory results and low cumulative cost (CHU et al., 1997), and is widely used in clinical practice, but it does not provide information on the intensity of inflammation. Due to the possibility of contamination by bacteria that also produce urease such as **Proteus sp** and **Pseudomonas sp**, leading to changes in the colour of the test during the process, it is advised that the preparation containing urea and the sensitive pH marker be done daily (NG, 1997).

Because it has been shown in the literature that the urease test is not 100% reliable, all the samples collected in this study were subjected to PCR (**Polymerase Chain Reaction**) to confirm the presence of the bacteria using specific **primers**, thus ruling out **"false negative"** results. **Of the 45 samples analysed, 14 (31%)** initially **tested** negative for the presence of the bacteria. However, PCR analysis revealed the presence of the bacteria, **confirming the "false negative" results for the** urease kit.

Analysing the samples showed that 21 patients (47%) had the **vacA-s1/m1** gene, of which 18 were positive for **cagA**. The **vacA-s1/m2** gene was also highly expressed in the samples and was found in 21 patients (47%), 19 of whom were positive for the **cagA** gene. Studies have shown that the vacA-s1/m1 gene is present in almost all strains of the bacterium and is largely responsible for the process of cytotoxin production in the body.

There is a growing body of literature suggesting that the **cagA** and **vacA** toxins interact and that this interaction has an effect on the severity of the disease. Initially, an antagonistic effect was demonstrated between these virulence factors, where **cagA** activates calcineurin through Cy phosphorolipase, while **vacA** inhibits this factor through calcium influx, preventing calcineurin activation. Another point demonstrated was that **cagA** not only blocks the cytotoxicity of **vacA**, but also prevents its ability to penetrate host cells (JONES et al., 2010).

The virulence factors **cagA** and **vacA** also show antagonistic activities in relation to cell morphology. In cells cultured on isogenic **H. pylori** mutant strains deficient in **cagA** or **vacA**, an increase in vacuolisation was observed in cells infected with **cagA** mutants, while cells infected with **vacA** mutants showed cell elongation. In short, the length was reduced in cells that had vacuoles and the number of vacuoles was reduced in elongated cells. There is an important **cagA** activation mechanism for cell spreading and morphological changes, while the **vacA** factor activates the inhibition of epidermal growth receptors. An explanation for this antagonism was suggested by AKADA et al., 2010, where the **cagA** factor is injected into cells where bacteria are associated, protecting the cells from the cytotoxic activity of **vacA**. The **vacA** then begins to attack the distant cells, releasing nutrients, and these interactions between the two strains combined may explain the observation of a link between the more active **vacA** allele (s1/m1), with the pathogenic **cagA** allele, manifesting more severe diseases (JONES et al., 2010).

Data from the National Cancer Institute 2012 indicate that despite being the second leading cause of cancer death in the world, in both sexes, **Helicobacter pylori** is showing a decline in its infection in most countries. Part of the explanation for this decline is due to factors related to the increased use of refrigerators for better food preservation, combined with changes in the population's eating habits (increased intake of fresh fruit and vegetables). Epidemiological studies suggest that a change in the population's dietary pattern, with an increased intake of fruit and vegetables, is associated with a lower risk of developing this neoplasm. The hypothesis that a healthy diet can be a protective factor is due to the fact that fresh fruit and vegetables contain vitamins with antioxidant properties, such as vitamins C and E and beta-carotene (INCA, 2012).

In general, environmental/behavioural factors are the main causes of gastric cancer. However, some studies show that genetic factors can influence the development of this neoplasm. One example is that the frequency of stomach cancer is approximately 20 per cent higher in individuals belonging to blood group A compared to other blood groups. This type of cancer does not have a good prognosis and the mortality/incidence ratio is considered high in all parts of the world. Its five-year relative survival is considered low in both developing and developed countries. Strategies for preventing stomach cancer include improvements in basic sanitation and changes in the population's lifestyle (INCA, 2012).

Another point to be discussed is the interaction between the bacteria and the host in relation to the receptors that the individual has, which is a considerable factor in the proliferation of the bacteria and the development of potential gastric diseases.

According to studies, **Helicobacter pylori** activates a transcription factor known as NF-kB which leads to the production of pro-inflammatory cytokines by gastric epithelial cells. However, the receptors for the bacteria's initial interaction with host cells that activate this signalling are not completely defined. It has been shown that microbial components that activate **Toll-Like** receptors (TLRs) lead to NF- kB dependent transcription and result in the production of pro-inflammatory cytokines. TLRs play a crucial role in the innate and adaptive immune response to microbial pathogens and their products. This receptor has a rich layer of leucine in its extracellular domains, similar to those of other pattern recognition proteins that promote binding. TLR proteins contain a cytoplasmic tail that is homologous to IL-1 and IL-18, which are receptors and can therefore trigger intracellular signalling pathways (SU et al., 2003).

Ten types of TLRs have been described, with TLR2 and TLR4 being the best characterised. TLR2 responds to peptidoglycan, lipoteichoic acid and bacterial lipoproteins. TLR4 is activated by lipopolysaccharide (LPS) from Gram-negative bacteria (such as **Helicobacter pylori)**. Studies have shown that TLR2 and TLR4 are expressed in human intestinal and epithelial cell lines (SU et al., 2003).

The adhesion of **Helicobacter pylori** bacteria to host epithelial cells is the first critical step in virulence. The microorganism binds to a variety of host cell molecules, including phosphatidylethanolamine, acetylneuraminylactose, GM3 ganglioside, sulphatides, lactotetraosylceramide, among others, which leads to its internalisation into gastric epithelial cells. It was shown that the TLR4 receptor on the cells increased the adherence of **Helicobacter pylori** compared to TLR2 receptor cells and with this it was observed that the expression of TLR4 is increased in gastric epithelia of infected individuals. This indicates that the exposed surface antigen is a biologically plausible receptor that is potentially available to mediate microbial adhesion. With this, we conclude that **Helicobacter pylori** infection induces the transcription and translation of TLR4 in a variety of gastrointestinal epithelial cells and when the bacterium is eradicated by bactericides, the receptors remain active, enabling and increasing the chances of new infections. These studies contribute to a better understanding of the cellular interactions between the bacteria and the host (SU et al., 2003).

Results were obtained for all the genes studied and for each pathological diagnosis the respective marker was shown, as shown in Graph 01.

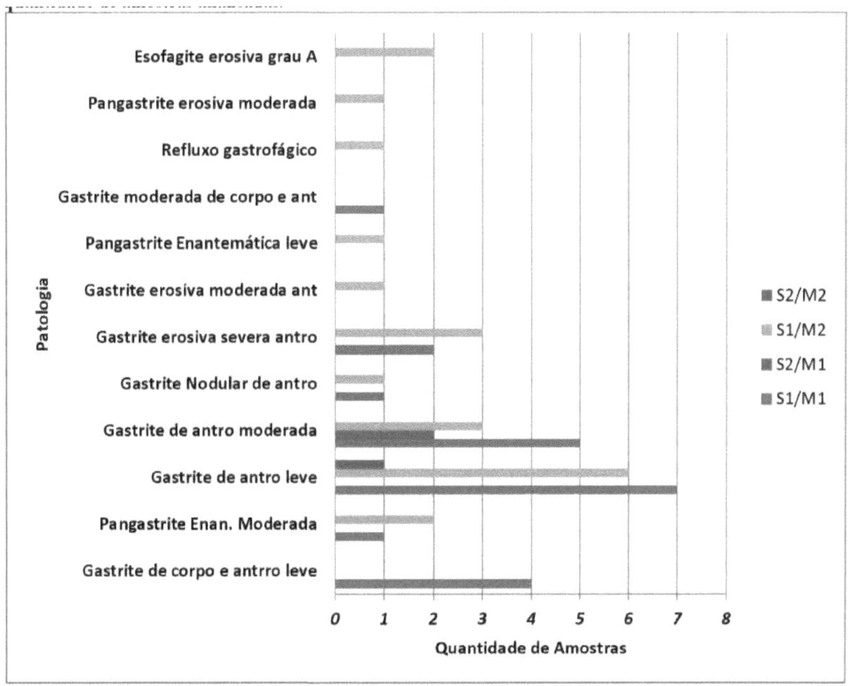

Graph 1: Demonstration of the pathologies and the respective genes that were expressed, according to the number of samples analysed.
Source: TOMANINI, R., 2012.

The results obtained point to a high incidence of contamination in patients by **Helicobacter pylori** at its most aggressive virulence level (**vacA-s1/m1** with **cagA+** and **vacA-s1/m2** with **cagA+**), thus confirming the literature presented in this study.

The experimental and clinical evidence that makes it possible to establish a pathogenic relationship between the presence of **Helicobacter pylori** and these gastrointestinal processes is presented in a study where the data collected is shown in Table 03 (FERNANDEZ, H., 2000).

Table 3: Occurrence of Helicobacter pylori **in different gastroduodenal pathologies.**

Patients	No. of studies reviewed	No. of patients studied	Presence of H. pylori %	
			Variation	Average
Chronic gastritis	15	1.411	62-94	83
Duodenal ulcer	12	620	60-100	92
Gastric ulcer	12	312	44-90	69
No dyspepsia Ulcer	09	1.293	33-87	51
Normal gastric histology	15	568	0-14	04

Source: FERNANDEZ, H., 2000.

The study proved the causal relationship of the bacterium with gastric pathologies, and two important observations were made: the direct conclusions indicate that experimental inoculation in human volunteers results in the production of gastritis; inoculation in experimental animals results in chronic gastritis in the inoculated animal; with appropriate antimicrobial therapy, in addition to eliminating the infection, regression of the gastritis is observed, returning the epithelium to normal. The indirect conclusions are that **H. pylori** only colonises the gastric epithelium and not other tissues; it is closely associated with epithelial cells of gastric origin, a fact rarely observed with epithelial cells of oesophageal origin or with gastric fibrolas; it is associated with certain types of gastroduodenal inflammation and not with all known types; it invariably triggers a specific systemic-humoral immune response. With antimicrobial therapy, the levels of specific anti-H.**pylori** antibodies decrease concomitantly with a reduction in inflammation, therapy with bismuth salts produces a reduction and elimination of inflammation and the bacterium is associated with epidemic and hypochlorhydric gastritis (FERNANDEZ, H., 2000).

The analysis carried out in this study found not only gastric pathologies in the aforementioned previous studies, but also other diseases directly related to infection by the bacteria, with the most prevalent being the more aggressive virulence factors reported in other studies.

Graph 01 shows the incidence of pathologies found in the analyses and their respective strains, which once again contributes to linking infection with various gastric diseases. It showed that of the 14 (fourteen) pathologies found, 07 (seven) were related to some type of gastritis, with only 01 (one) case of moderate erosive gastritis of the antrum not showing the **vacA s1/m1** strain, unlike the other cases in which all the pathologies showed this virulence. It was also found that, like the **s1/m1** strain, the **s1/m2** strain was not diagnosed in only 2 (two) cases of gastritis. The diagnoses presented for gastritis included mild antrum gastritis with seven (07) samples containing the **s1/m1** strain and six (06) samples containing the **s1/m2** strain, and moderate antrum gastritis with five (05) samples containing the **s1/m1** strain, two (02) samples containing the **s2/m1** strain and three (03) samples containing the **s1/m2** strain.

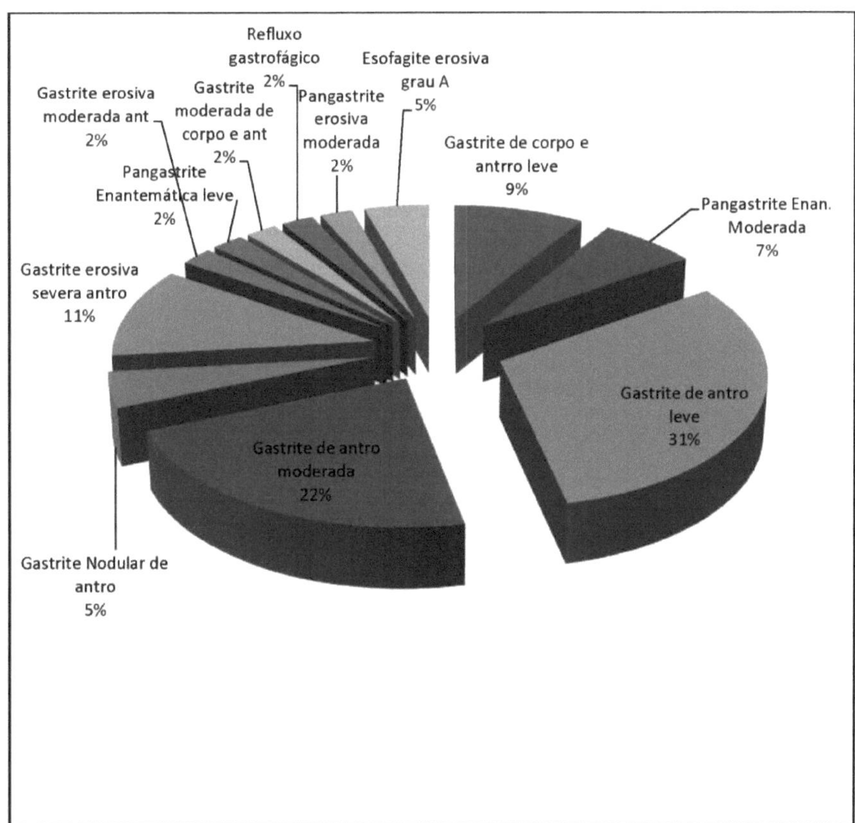

Graph 02: Percentage of pathologies in the samples
Source: TOMANINI, R., 2012.

The incidence of diseases such as gastritis in cases of **Helicobacter pylori** infection in the samples analysed is noteworthy; graph 02 shows that more than 50% of the cases were gastritis. This rate corroborates the literature already presented, showing that when it comes to infection with the bacterium, the chances of developing this disease are high and in most cases the treatment can be simple and effective when diagnosed early and correctly.

It is worth noting that although the percentage of diagnoses for pangastritis was only 9% of the cases analysed, this type of pathology requires special attention because it involves a large part of the stomach. In this case, there is an increased chance of more serious illnesses developing, and immediate treatment is essential when a positive diagnosis is made.

Helicobacter pylori is a bacterium with a worldwide distribution that has only been isolated from the gastroduodenal epithelium of humans. There are no known animal or environmental reservoirs for this bacterium, and it is postulated that the most common mode of transmission could be from person to person and the route of infection, the oral route. Once acquired, the infection can persist for years and even for life.

Although the most common source of transmission is person-to-person, iatrogenic infections, derived from the use of contaminated instruments, have been described (FERNANDEZ, H., 2000). Another point to highlight in the infectious process is the inadequate socio-economic and environmental conditions in certain regions, leading to a lack of information about personal hygiene care and precarious public health conditions for the population.

Various ways of preventing infection by microorganisms are publicised daily by the media in order to avoid the development of diseases and their transmission. Bacteria, viruses and fungi are the main agents of countless diseases that develop in the human organism. Through simple means, they penetrate the human body and find satisfactory conditions to develop and reproduce. As already mentioned, not only unhealthy or genetic factors influence the infection and development of microorganisms, but also the individual's quality of life is a fundamental factor.

With technological advances in medicine, previously untreatable illnesses that caused people to suffer for a long time and even die are now treated simply and effectively. However, there is still a great deal of human resistance to prophylactic health measures, which means that people only go to the doctor when symptoms appear, and in some cases the disease is irreversible. It can be seen from this that the oldest and simplest remedy against disease is prevention and that although it has been a widespread thesis for many years, most human beings have still not learnt their lesson.

CHAPTER 5

Bibliographical references

AKADA, J.K; AOKI, H.; TORIGOE, Y.; KITAGAWA, T.; KURAZONO, H.; HOSHIDA, H.; NISHIKAWA, J.; TERAI, S.; MATSUZAKI, M.; HIRAYAMA, T.; NAKAZAWA, T.; AKADA, R.; NAKAMURA, K. Helicobacter pylori cagA **inhibits endosytosis of cytotoxin** vacA **in host cells.** Dis. Model. Mech. 3, 305-607, 2010.

ATHERTON, J.C.; BLASER, M.J. Coadaptation of **Helicobacter pylori** and humans: ancient history, modern implications. **J.C.I.**, 119(9): 2475-2487, 2009.

ATHERTON, J.C.; CAO, P.; COVER, T.L. Mosaicism in vacuolating cytotoxin alleles of **Helicobacter pylori**: association of especific **vacA** types with cytotoxin production and peptic ulceration. **J. Biol. Chem.**, 270:17771-7, 1995.

ARENTS, N.L.; VAN ZWET, A.A.; THIJS, J.C.; KOOISTRA-SMID, A.M.; VAN, S.; DEGENER, J.E.; KLEIBEUKER, J.H.; VAN DOORN, L.J. The importance of **vacA**, **cagA**, and **iceA** genotypes of **Helicobacter pylori** infection in peptic ulcer disease and gastroesophageal reflux disease. **Am J. Gastroenterol**, 96(9):2603-8, 2001.

BETTEN, A.; BYLUND, J.; CHRISTOPHE, T.; BOULAY, F.; ROMERO, A.; HELLSTRAND, K.; DAHLGREN, C. A proinflammatory **pectide** from **Helicobacter pylori** activates monocytes to induce lymphocyte dysfunction and apoptosis. **J. Clin. Invest.**, 108(8):1221-8, 2001.

BLASER, J.M. **Helicobacter pylori**: its role in disease. **Clin. Infect. Dis.**, 15:386-393, 1992.

BLASER, J.M.; ATHERTON, J.C. **Helicobacter pylori** persistence: biology and disease. **J. Clin. Invest.** 2004.

BRANSON, D. **Methods in Clinical Bacteriology - A Manual of Tests and Procedures.** Springfield, III; 1973.

BROOKS, E.G.; BAMFORD, K.B.; DENNING, T.L.; PAPPO, J.; ERNST, P.B. Negative Selection of T Cells by **Helicobacter pylori** as a Model for Bacterial Strain Selection by Immune Evasion. **The J. of Immunology**, 167:926-934, 2000.

BRUNTON, L.L; LAZO, J.S.; PARKER, K.L. Goodman & Gilman: **The Pharmacological Basis of**

Therapeutics. 11 ed. Rio de Janeiro: McGaw-Hill, p. 9950 and 880, 2006.

BRZOZOWSKI, T. et al. Mucosal irritation, adaptive cytoprotection, and adaptation to topical ammonia in the rat stomach. **Scand. J. Gastroenterol**. v.31, n.9, p. 837-46, 1996.

CHEY, W.D.; WONG, B.C. Practice Parameters Committee of the American College of Gastroenterology. American College of Gastroenterology Guideline on the Management of **Helicobacter pylori** Infection. **Am J Gastroenterol**; v.102, n. 8, p.1808-25, 2007.

CHOLI-PAPADOPOULOUS, T.; KOTTAKIS, F.; PAPADOPOULOU, G.; PENDAS, S. **Helicobacter pylori** neutrophin activating protein as target for new drugs against **H. pylori** inflammation. **World J. Gastroenterol**, 17(21):2585-2591, 2011.

CHU, K.M.; POON, R.; TUEN, H.H.; LAW, S.Y.K.; BRANICKI, F.J.; WONG, J. A prospective comparison of locally made rapid urease test and histology for the diagnosis of **Helicobacter pylori** infection. **Gastrointest Endosc**. 46: 503, 1997.

COTRAN, R.S.; KUMAR, V.; COLLINS, T. R. **Structural and Functional Pathology**. 6ed. Rio de Janeiro: Guanabara Koogan, p. 283 and 708 - 721, 2000.

COVER, T.L.; TUMMURU, M.K.; CAO, P.; THOMPSON, S.A.; BLASER, M.J. Divergence of genetic sequences for the vacuolating cytotoxin among **Helicobacter pylori** strains. **J. Biol. Chem.**, 269:10566-73, 1994.

COVER, T.L.; HANSON, P.I.; HEUSER, J.E. Acid-induced dissociation of **vacA**, the **Helicobacter pylori** cytotoxin, reveals its pattern of assembly. **J. Cell. Biol**. 138:759-769, 1997.

CURRENT EUROPEAN. **Concepts in the management of** Helicobacter pylori **infection.** The Maastricht Consensus Report. European **Helicobacter Pylori** Study Group. Gut; v.41, n.1, p.8-13, 1997.

DING, S.; GOLDBERG, J. B.; HATAKEYAMA, M. **Helicobacter pylori** infection, oncogenic pathways and epigenetic mechanisms in gastric carcinogenesis. **Future Oncol.**, 6(5):854-862, 2010.

DIXON, M.F.; GENTA, R.M., YARDLEY, J.H.; CORREA, P. Classification and grading of gastritis. The updated Sydney System. International Workshop on the Histopathology of Gastritis, Houston 1994. **Am J Surg**

Pathrol, 1996.

DORER, M.S.;FERO, J.; SALAMA, N.R. DNA damage triggers genetic Exchange in **Helicobacter pylori**. **Plos Pathog**. 6(7): 1-10, 2010.

EATON, K.A.; MORGAN, D.R.; KRAKOWA, S. **Campylobacter pylori** virulence factors in genobiotic piglets. **Infec. Immun.**, 57:1119-25, 1989.

EATON, K.A.; KRAKOWKA, S. Arvirulent urease-deficient **Helicobacter pylori** colonises gastric epithelian explants ex vivo. **Scand J. Gastroenterol**, 30:434-37, 1995.

EISIG, N.J.; CARVALHAES, A. **Peptic ulcer and** H. pylori. Moreira Jr. Editora, São Paulo, 2006.

ERNST, P.B.; PUERA, D.A.; CROWE, S.E. The translation of **Helicobacter pylori** basic research to patient care. **Gastroenterol**, 130(1): 188-206, 2006.

ESKANDAR, K.S.; ABDULAH, B.; MALIHE, M.; KAMRAN, L.; ALI-REZA, T.; MEHDI, S.R. Association of **H. pylori cagA** and **vacA** genotypes and IL-8 gene polymorphisms with clinical outcome of infection in Iranian patients with gatrointestinal diseases. World **J. Gastrointestinal**, 12(32): 5205-5210, 2006.

FERNANDEZ, HERIBERTO. Genus **Helicobacter in** TRABULSI, L.R.; ALTERTHUM, F.; GOMPERTZ, O.F.; CANDEIAS, J.A.N. **Microbiologia.** 3ª ed. São Paulo: Atheneu, P.263-267 , 2000.

GOBERT, A.P.; MERSEY, B.D.; CHENG, Y.; BLUMBERG, D.R.; NEWTON, J.C.; WILSON, K.T. Cutting edge: urease release by **Helicobacter pylori** stimulates macrophage inducible nitric oxide synthase. **J. Immunol**. 168: 6002-6, 2002.

GOODMAN, K.J.; CORREA, P.; MERA, R.; YEPEZ, M.C.; CERÓN, C.; CAMPOS, C.; GUERRERO, N.; SIERRA, M.S.; BRAVO, L.E. **Effect of** Helicobacter pylori **Infection on Growth Velocity of School-age Andean Children.** Epidemiology. Author manuscript; available in PMC, 2011.

GRAHAM, D.Y. Therapy of **Helicobacter pylori**: current status and issues. **Gastroenterology**, 118:S2-S8, 2000.

GRAHAM, D.Y. Antibiotic resistance in **Helicobacter pylori**: implication for therapy. **Gastroenterology**. v.115,

p. 1272-7, 1998.

GUNN, M.C.; STEPHENS, J.C.; STEWART, J.A.; RATHBONE, B.J.; WEST, K.P. The significance of **cagA** and **vacA** subtypes of **Helicobacter pylori** in the pathogenesis of inflammation and peptic ulceration. **J. Clin. Pathol.**, 51(10):761-4, 1998.

HANCOCK, R.E.; ALM, R.; BINA, J.; TRUST, T. **Helicobacter pylori**: a surprisingly conserved bacterium. **Nat. Biotechnol**, 16:216-7, 1998.

HATAKEYAMA, M. Oncogenic machanisms of the **Helicobacter pylori cagA** protein. **Nat. Rev. Cancer**, 4(9):688-94, 2004.

HATAKEYAMA, M. The interaction between cagA-positive **Helicobacter pylori** and gastric epithelial cells. **Nature Reviews Cancer 4**, 688-694, 2004.

HEATLEY, R.V. **The *Helicobacter pylori* handbook.** 1ª Ed. Osney Mead. Balckwell Science, 1995.

HELICOBACTER PYLORI. Available at: <http://bioweb.uwlax.edu>. Accessed on 22/May/2012.

HELICOBACTERFOUDATION . Available at : <http://www.helico.eom/h epidemiology.html>. Accessed on: 22/May/2012.

INCA, Coordination of Cancer Prevention and Surveillance. **Cancer Incidence in the World**. Rio de Janeiro: INCA, 2003.

INCA, Coordination of Cancer Prevention and Surveillance. **Estimates 2008:** Cancer Incidence in Brazil. Rio de Janeiro: INCA, 2007.

INCA, Coordination of Cancer Prevention and Surveillance. **Estimate 2012:** Cancer incidence in Brazil. Rio de Janeiro: INCA, 2012.

ISOMOTO, H.; MOSS, J.; HIRAYAMA, T. Pleiotropic actions of *Helicobacter pylori* vacuolating cytotoxin, *vacA*. **Tohoku J. Exp. Med.**, 220:3-14, 2010.

JENKS, P.J.; KUSTERS, J.G. Pathogeneses and virulence of *Helicobacter pylori*. **Curr. Opin. Gastroenterol**, 16(suppl.1):S11-8, 2000.

JONES, K.R.; WHITMIRE, J.M.; MERRELL, D.S. A tele of two toxins: *Helicobacter pylori cagA* and *vacA* modulate host pathways that impact disease. **Forntiers in Microbiology**. 1:115, 2010.

KELLEY, J.R.; DUGGAN, J.M. Gastric cancer epidemiology and risk factor. **J. Clin Epidemiol**, 56:1-9, 2003.

KENNEMANN, L.; DIDELOT, X.; AEBISCHER, T.; KUHN, S.; DRESCHER, B.;DROEGE, B.; REINHARDT, M.; CORREA, P.; MEYER, T.F.; JOSENHANS, C.; FALUSH, D.; SUERBAUM, S. **Helicobacter pylori** genome evolution during human infection. **P.N.A.S.**, 108(12): 5033-5038, 2011.

KODAIRA, M.S.; ESCOBAR, A.M.U. Aspectos epidemiológicos do *Helicobacter pylori* na infância e adolescência. **Rev. Saúde Pública**, 2002.

KONTUREK, J.W. Discovery by Jawordki of *Helicobacter pylori* and its pathogenetic role in peptic ulcer, gastritis and gastric cancer. **J. Phisiol Pharmacol**, 54 (Suppl3):23-41, 2003.

KORWIN, J.D. **Faut-il éradiquer** Helicobacter pylori **en cas de pathologie de reflux ?** Service de Médecine Interne H, CHU de Nancy, FRANCE, vol. 30, n° 26, pp. 1313-1320, 2001.

KOSTRZYNSKA, M.; BETTS, J.D.; AUSTIN, J.W.; TRUST, T.J. Identification, characterisation, and spatial localization of two flagellin spicies in *Helicobacter pylori* flagella. **J. Bacteriol**, 173:937-46, 1991.

KRAKOWKA, S.; EATON, K.A.; LEUNK, R.D. Antimicrobial Therapies for *Helicobacter pylori* Infection in Gnotobiotic Piglets. **Antimicrob Agents Chemother**. 42(7):1549-1554, 1998.

KUSTERS, J.G.; GERRITS, M.M.; VAN, J.A.S.; VANDENBROUCKE-GRAULS, C.M. Coccoid forms of *Helicobacter pylori* are the morphologic manifestation of cell death. **Infect Immun**, 65:3672-9, 1997.

LEYING, H; SUERBAUM, S.; GEIS, G. HAAS, R. Cloning and genetic characterisation of *Helicobacter pylori* flagellin gene. **Mol. Microbiol**, 6:2863-74, 1992.

LIND, T.; MEGRAUD, F.; UNGE, P.; BAYERDORFFER, E.; O'MORAIN, C.; SPILLER, R.; ZANTEN, S.V.V.;

BARDHAN, K.D.; HELLBLOM, M.; WRANGSTADH, M.;

ZEIJLON, L.; CEDERBERG, C. The MACH2 study: Role of omeprazole in eradication of **Helicobacter pylori** with 1-week triple therapies. **Inst. of Gastroenterology**, 1999.

LUPETTI, P.; HEUSER, J.E.; MANETTI, R.; MASSARI, P.; LANZAVECCHIA, S.; BELLON, P.L.; DALLAI, R.; RAPPUOLI, R.; TELFORD, J.L. Oligomeric and subunit structure of the **Helicobacter pylori** vacuolating cytotoxin. **J. Cell. Biol.**, 133(4):801-7, 1996.

MALFERTHEINER, P. et al. **European Helicobacter pylori study group EHPSG. Current concepts in the management of Helicobacter pylori infection**. The maastricht 22000 Consensus report. Aliment Pharmacol Ther; v.16, n.2, p.167-80, 2002.

MALFERTHEINER, P. et al. **Current concepts in the management of** Helicobacter pylori **infection - The Maastricht III Consensus Report**. Gut. v.56, n.6, p.772-81, 2007.

MARSHALL, B.J. Virulence and pathogenicity of *Helicobacter pylori*. **Journal of Gastroenterology and Hepatology**; v.6, p.121-124, 1991.

MARSHALL, B.J; WARREN, J.R. **Unidentified curved bacilli in the stomach of patients with gastritis and peptic ulceration.** Lancet. v.1, n.8390, p.1311-5, 1984.

MEGRAUD, F. et al. *Helicobacterpylori* resistance to antimicrobial agents after failure of an initial treatment and impact on results of second-line treatment strategies: A multicenter prospective study. **Gastroenterol**; v.120(supl.1), p.A15, 2001.

MITANI-EHARA, S. **Studies on gastric mucosal cell injury induced by** Helicobacter pylori. Hokkaido Igaku Zasshi. v.69, n.4, p.836-46, 1994.

MITANI-EHARA, S. et al. Studies on gastric mucosal cell injury induced by *Helicobacter pylori*. **J. Clin. Gastroenterol.** v.25, p. S164-8, 1997.

MOBLEY, H.L.T; MENDZ, G.L; HAZELL, S.L. *Helicobacter pylori*: physiology and genetics. **Washington: ASM Press**, p. 03-91 and 459-530, 2001.

MONTECUCCO, C.; PAPINI, E.; DE BERNARD, M.; ZORATTI, M. Molecular and cellular activities of *Helicobacter pylori* pathogenic factors. **FEBS Lett**, 452(1-2):16-21, 1999.

MONTECCUCO, C.; RAPPUOLI, R. Living dangerously: how *Helicobacter pylori* survives the human stomach. **Nature Rev.**, 2:457-66, 2001.

MORAES-FILHO, J.P.P.; CECCONELLO, I.; RODRIGUES, J.G.; CASTRO, L.P.; HENRY, M.A.; MENEGHELLI, U.G.; QUIGLEY, E. Brazilian Consensus on Gastroesophageal Reflux Disease: Proposals for Assessment, Classification and Management. **Am J Gastroenteral**, 97:241-248; 2002.

MURRAY, P.R. et al. **Manual of Clinical Microbiology**. 9ed. v.1, p. 947-956, 2007.

NAITO, Y.; YOSHIKAWA, T. Molecular and cellular mechanisms involved in *Helicobacter pylori* induced inflammation and oxidative stress. **Free Radic Biol. Med.**, 33(3):323-36, 2002.

NG, F.H.; WONG, S.Y.; NG, W.F. Storage temperature of the unbuffered rapid urease test. **Am J Gastroenterol**. 92: 2230, 1997.

OLIVEIRA, A.M.; ROCHA G.A.; QUEIROZ, D.M.; DE MOURA, S.B.; RABELLO, A.L. Seroconversion for *Helicobacter pylori* in adults from Brazil. **Trans. R. Soc. Trop. Med. Hyg**, 93:261-3, 1999.

ORNELLAS, L.C.; CURY, M. S.; LIMA, V.M.; FERRARI, A.P. Evaluation of the rapid urease test preserved in the refrigerator. **Arq Gastroenterol**. 37:1555-157, 2000.

PAGLIACCIA, C.; DE BERNARD, M.; LUPETTI, P.; JI, X.; BURRONI, D.; COVER, T.L.; PAPINI, E.; RAPPUOL, R.; TELFORD, J.L.; REYRAT, J.M. **The m2 form of the** Helicobacter pylori **has cell type-specific vacuolating activity.** Proc. Natl. Acad. Sci. USA. 95(17): 10212-7, 1998.

PAPINI, E.; ZORATTI, M.; COVER, T.L. In search of the **Helicobacter pylori vacA** mechanism of action. **Toxicon**, 39(11):1757-67, 2001.

PEEK Jr., R.M.; CRABTREE, J.E. **Helicobacter** infection and gastric neoplasia. **J. Phatol**, 208(2):233-48, 2006.

PETERSON, W.L.; GRAHAM, D.Y. *Helicobacter pylori* : **Sleisenger and Fordtran's gastrointestinal and liver**

disease: pathophisiology, diagnosis, management. 7ª edition. Philadelphia: Ed. Saunders, Cap 39 - pag. 732-46, 2002.

POUNDER, R.E.; LEWIN, J.; DHILLON, A.P.; SIM, R.; MAZURE, G.; WAKEFIELD, A.J. Persistent measles virus infection of the intestine: confirmation by immunogo electron microscopy. **Intern. J. Gastr. Hepat.**, 1995.

ROBERTSON, M.S.; CLANCY, R.L.; CADE, J.F. **Helicobacter pylori** in intensive care: why we should be interested. **Intensive Care Med**, 29:1881-8, 2003.

RODRIGUES, R.T.; BRAGA, L.L. Prevalence of **Helicobacter pylori** infection in children from an urban community in north-east Brazil and risk factors for infection. **Eur. J. Gastroenterol Hepatol.**, 16:201-5, 2004.

SEPÚLVEDA, A.R. Molecular testing of **Helicobacter pylori** associated chronic gastritis and premalignant gastric lesions. **J.Clin. Gastroenterol**, 2001.

SIQUEIRA, J.S.; et al. **General Aspects of *Helicobacter pylori* Infections.** Review. RBAC. v.39, n.1, p.9-13, 2007.

SMOOT, D.T. et al. **Helicobacter pylori** urease activity is toxic to human gastric epithelial cells. **Infect Immun.** v.58, n.6, p.1992-4, 1990.

SOUTO, F.J.; FONTES, C.J.; ROCHA, G.A.; DE OLIVEIRA, A.M.; MENDES, E.N.; QUEIROZ, D.M. **Prevalence of *Helicobacter pylori* infection in a rural area of the state of Mato Grasso, Brazil.** Mem. Inst. Oswaldo Cruz, 93:171-4, 1998.

SOUZA, R.C.; LIMA, J.H. **Helicobacter pylori** and gastroesophageal reflux disease: a review of this intriguing relationship. **Dis. Esophagus**, 22(3):256-63, 2009.

STOICOV, C.; WHARY, M.; ROGERS, A.B.; LEE, F.S.; KLUCEVSEK, K.; LI, H.C. Coinfection modulates inflammatory responses and clinical outcome of **Helicobacter pylori felis** and toxoplasma gondii infection. **J. Immunol.**, 173:3329-36, 2004.

SU, B.; CEPONIS, P.J.M.; LEBEL, S.; HUYNH, H.; SHERMAN, P.M. **Helicobacter pylori** Toll-Like receptor 4 expression in gastrointestinal epithelial cells. **Infection and Immunity**, 71(6):3496-3502, 2003.

SUERBAUM, S.; JOSENHANS, C.; LABIGNE, A. Cloning and genetic characterisation of the **Helicobacter pylori** and **Helicobacter mustelae flab** flagellin genes and construction of **H.** *pylori flaA* - and **fab-negative** mutants by electroporations-mediated allelic exchange. **J. Bacteriol**, 175:3278-88, 1993.

TOMBOLA, F.; MORBIATO, L.; DEL GUIDICE, G.; RAPPUOLI, R.; ZONATTI, M.; PAPINI, E. The *Helicobacter pylori vacA* toxin is a urea permease diffusing across epithelia. **J. Clin. Invest.**, 108:929-937, 2001.

TRABULSI, L.R.; ALTERTHUM, F.; GOMPERTZ, O.F.; CANDEIAS, J.A.N. **Microbiologia.** 3ª ed. São Paulo: Atheneu, P. 92, 97, 103-109, 264, 266 and 291, 2000.

VAN DOORN, L.J.; FIGUEIREDO, C.; SANNA, R.; PLAISIER, A.; SCHNEEBERGER, P.M.; BOER, W.; QUINT, W. Relevance of the *cagA*, vacA, and *iceA* status of *Helicobacter pylori*. **Gastroenterology**, 115:58-66, 1998.

VAN DOORN, L.J.; SCHNEEBERGER, P.M.; NOUHAN, N; PLASIER, A.; QUINT, W.; BOER, W. Importance of *Helicobacter pylori cagA* and *vacA* status for the efficacy of antibiotic treatment. **Gastroenterology**, 46:321-6, 2000.

VOLAND, P.; WEEKS, D.L.; MARCUS, E.A.; PRINZ, C.; SACHS, G.; SCOTT, D. Interactions among the seven **Helicobacter pylori** proteins encoded by the urease gene cluster. **Am J Physiol Gastrointest Liver Physiol**, 284:G96-106, 2003.

WANG, H.J.; KUO, C.H.; YEH, A.A.; CHANG, P.C.; WANG, W.C. Vacuolating toxin production in clinical isolates of **Helicobacter pylori** with different **vacA** genotypes. **J. Infect. Dis.**, 178(1):207-12, 1998.

WEEKS, D.L.; SACHS, G. Sites of pH regulation of the urea channel of **Helicobacter pylori**. **Mol. Microbiol.**, 40(6):1249-89, 2001.

YAMAOKA, Y.; Mechanisms of disease: **Helicobacter pylori** virulence factors. **Nat. Rev. Gastroenterol Hepatol**, 7(11): 629-41, 2010.

ANNEXES

Annex I

DNA extraction

1) Place the biopsy in a falcon

2) Add 300µl of digestion buffer

3) Add a grain of proteinase K

4) Incubate at 37° C for at least 5 hours or overnight

5) Add 400µl of ice-cold PCI (phenol: chloroform: isoamyl = 25:24:1)

6) Mix by inversion (1 min). Do not use cortex

7) **Centrifuge for 5 ' rotation (13 .OOOrpms)**

8) Transfer the DNA (supernatant) to a new tube

9) Add 400µl of ice-cold CI (Chloroform: 1900Mol(a) Cohal 24:1)

10) Mix by inversion for 1 minute (without vortexing)

11) Centrifuge for 5 min. rotation (13,000rpms)

12) Transfer the DNA (supernatant) to a new tube

13) Repeat the previous steps (from item no. 5)

14) Additional 600µl isopropanol and 100µl 3Mol NaAc (sodium acetate) (room temp)

15) Mix carefully by hand

16) Do not fish out the DNA (for biopsy). Centrifuge for 10 min. rotation (13,000rpms). Discard in sink (or remove liquid with pipette).

17) Wash the DNA with 250 µl of ice-cold 70% ethanol (-20° C). Centrifuge and discard

18) Incubate the samples at 37° C for 10-15 min. With lid open

19) Add 100-200 µl of myeli-Q H2O

20) Leave the sample at 37°C or room temperature for 15 min.

21) Check result on spectrophotometer

22) Check DNA with 0.8% agarose gel

Annex II

PCR - Hpnap

- **H2O=**   16.00µl

- Buffer   = 2,50µl

- **MgCl2=**   0.75µl

- Primer 1   = 1,25µl

- Primer 2   =1,25µl

- **Taq=**   0.25µl

- **DNA=**   2.5µl

- Final Value = 25,00µl

Agarose gel electrophoresis (0.8%)

- For 100mL of gel, weigh out 0.8g of agarose
- Add 100mL of TAE 1X
- Microwave until smooth
- Prepare the electrophoresis vat with a comb compatible with the number of samples, including the control sample and the blank.
- After homogenising the agarose, pour it into the vat and wait for approximately 15 minutes for the gel to solidify.
- Add 4µl of bromophenol blue to each sample
- After the gel has solidified, top up the gel with 5X TBE buffer
- Place 8µl **of 100pb ladder** in the first well
- Place 10µl of each sample, including the control and blank sample, in the subsequent wells
- Switch on the equipment at 100V, 110M.A for 1 hour.

I want morebooks!

Buy your books fast and straightforward online - at one of world's fastest growing online book stores! Environmentally sound due to Print-on-Demand technologies.

Buy your books online at
www.morebooks.shop

Kaufen Sie Ihre Bücher schnell und unkompliziert online – auf einer der am schnellsten wachsenden Buchhandelsplattformen weltweit! Dank Print-On-Demand umwelt- und ressourcenschonend produziert.

Bücher schneller online kaufen
www.morebooks.shop

Printed by Books on Demand GmbH, Norderstedt / Germany